THE SELECTION OF DESIGN

THE SELECTION OF
DESIGN

GORDON L. GLEGG

CAMBRIDGE
AT THE UNIVERSITY PRESS
1972

Published by the Syndics of the Cambridge University Press
Bentley House, 200 Euston Road, London NW1 2DB
American Branch: 32 East 57th Street, New York, N.Y. 10022

© Cambridge University Press 1972

Library of Congress Catalogue Card Number: 72-80591

ISBN: 0 521 08686 8

Printed in Great Britain
at the University Printing House, Cambridge
(Brooke Crutchley, University Printer)

CONTENTS

INTRODUCTION

Clothed in design, the adventure of invention should both fascinate the engineer and benefit the community. Usually, adventure is a strictly personal affair, and so is inventing, for both are often rooted in the unpredictable. Many inventors are unpredictable, in more ways than one, but others find inspiration within a more ordered context. In fact inventors can be, to a certain extent, divided into two different types.

There are those who are intuitive, whose ideas fall like a bolt from the blue and who find little help in formalised procedure.

In contrast, others think most creatively within a planned framework and, although personal inspiration is still always useful and sometimes essential, it must, for them, be within systematic thought.

Especially at the beginning of their careers, when experience or self-confidence is lacking, young inventors tend to be of the latter type and welcome a systematic approach to engineering problems. This book is intended to guide them in this. It suggests how to select where to begin in inventing, how to select designs and how to select the best of them. These systematic patterns of thought are intended to apply over the whole spectrum of design, not just a local technology, and enable one at least to make a start at tackling a problem and often, I hope, evolve a successful conclusion. And the process of doing so need not be as dull as it sounds.

1

SELECTING A BEGINNING

The modern engineer is rarely faced with a simple problem. Most of the simple ones have been solved already. In earlier days each step in a process would be done by a separate machine or each force in a structure well smothered by redundant members.

Today's relentless economic pressures dictate the use of multi-stage machines with intricate controlling systems to lift speed and precision beyond anything the human hand or eye could ever approach. There is no unsophisticated engineering left; or there shouldn't be.

Our grandfathers, and even our fathers, could visualise and invent total machines. Today we usually cannot hold in our mind's eye the total requirements of a design, much less how to achieve them.

Where do we start? Where, so to speak, does the vital centre of design gravity lie? If we try to start everywhere we get nowhere. We must decide which is the horse and which the cart, otherwise we may find the horse asleep in it or under it. If there is any general answer to this problem, perhaps we may find it by examining where, in fact, are the points where successful designers have chosen to begin over a wide field and see if any common strategy emerges.

We will first have a look at one of the commonest machines in industrial use, i.e. the production line where materials are put in at one end and a saleable product comes out at the other. There are hundreds of different varieties of initial materials and final products but the machines that link them have all, or nearly all, some common characteristics.

One is that they are all intended to run at a profit. This means that the money you obtain for the material in the form that it comes out of the machine must be quite a deal more than the cost of buying the same materials to put into the machine. Usually there is a variety of raw materials that have to be fed in and subsequently mixed up together in rigidly controlled proportions. Especially in exploiting plastics, some of the materials will be liquids and others ground-up solids. A constant stream of lorries arrives each day to transfer their contents to tanks or hoppers with the minimum of delay. And how can you be sure that you are not being swindled? Only by having a quick method of checking the amount of material being delivered. And the only quick way is to do it by weight; weigh the vehicle when full, re-weigh when empty, and there is no argument, or there ought not to be. More often than not it is the only way you can do it in any case. If the material arrives in irregular slabs or bars you cannot measure up each. Even more impossible is the situation with granules or powders. You cannot assess individual grains, and their total bulk volume depends largely on the amount of entrained air. In general, then, raw materials have either unpredictable shapes or volumes or both. And so they must be metered by weight. Volume is largely irrelevant and wholly unobtainable. And after the materials have been put into the production line the same principle applies. All control must be done by weight. If you try to do it by volume, variations in bulk density will produce absolute chaos.

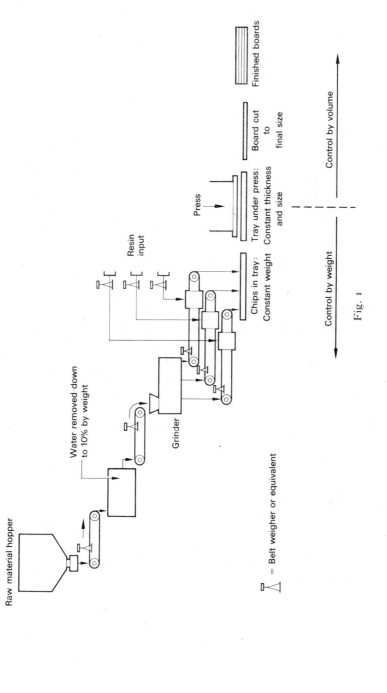

Raw material hopper

Water removed down to 10% by weight

Grinder

Resin input

Chips in tray: Constant weight

Press

Tray under press: Constant thickness and size

Board cut to final size

Finished boards

△▷ = Belt weigher or equivalent

Control by weight

Control by volume

Fig. 1

Now look at the other end of the machine. Saleable products are coming out: floor coverings, washing-up bowls, strips, sheets or shapes, and the customer is largely, and usually totally, concerned with their volume. Weight is irrelevant. The housewife who wants a new floor covering for the kitchen is concerned with its thickness and the price per square yard. And this is but one among thousands of possible illustrations. Volume, or volume in a certain shape, is what the market requires.

In short, the manufacturer buys his materials by weight and then sells them by volume. He hasn't any option.

And now we must look in more detail at his production line. At the beginning of it all control must be done by weight; at the end weight is irrelevant and volume vital; and so all control must be by volume. And somewhere in the middle is a frontier where 'control by weight' hands its product over to 'control by volume'. And what happens at the frontier crossing is vital. How are you going to form your material? Everything before the frontier must be designed in anticipation of this, everything on the further side as a result of it. And this is where your designing must begin. Find the frontier where forming energy is fed in (rolls, presses, etc.), and design backwards and forwards from it. Fig. 1 shows a simplified outline of a chip-board production line and the position of the frontier.

But is there a possibility that there is more than one frontier in a design? There certainly is, and much trouble can arise if it is overlooked.

The comparative evolution in design of the motor car engine and the motor cycle engine may illustrate this.

6

From the point of view of mechanical efficiency early motor cycle engines were designed in the right way for the wrong reasons, while those of cars were designed in the wrong way for the wrong reasons. I remember in 1950 listening to a car salesman who was eulogising about his latest model. He said (quite correctly) that it had a 'square engine', i.e. the bore and stroke of its cylinders were almost equal. The actual dimensions were a bore of 84 mm and a stroke of 90 mm. Again quite correctly, he said that all motor manufacturers would copy this feature in future and the 'square engine' would become the orthodox design. Then he spoilt it all by saying that the idea was entirely new and the greatest discovery since the jet engine. I did not think it would be polite to tell him that I bought a twin cylinder J.A.P. motor cycle engine, whose bore and stroke were equal (both 84 mm), very second hand in 1929, when it had already been a most popular design for five years.

Thermo-dynamically remarkably efficient, it was in quantity production and used in the Brough Superior Motor Bicycle, each of which was sold with a signed certificate that it had already exceeded 100 m.p.h. over a timed $\frac{1}{4}$ mile. These timed runs were made on a narrow private road only $1\frac{3}{4}$ miles long.

The quarter of a century lag in design efficiency was due to where the designers had selected their starting points. For the first half of this century motor cars were taxed according to the square of the bore of their cylinders. You could have any stroke you liked free of charge. And so manufacturers elongated the stroke as far as they could, which limited the maximum r.p.m. and left no room for efficient valve sizes. The design started in a Government department and ended in a mess.

7

In the case of the motor bicycle you paid tax because you had one but no one worried about its engine. Especially if it was a twin cylinder machine everything tended to make you want to keep the cylinders as short as possible. You felt it would be wise to have at least a little clearance between the almost red-hot cylinder head and the bottom of the petrol tank. Also a lower centre of gravity meant that it was less likely to fall over, hurt you less when it did, and was easier to pick up when it had. As the width between the frame members was also limited a 'square' engine seemed the best idea.

This basic divergence in technique has now disappeared, for, except in unusual and exceptional circumstances, we design the cylinder head first in both instances and go forward and backwards from there. The frontier lies through the combustion chamber, for it is there that the air–petrol mixture passes from a passive to an active role, where power springs out of almost nothing.

But the importance of this can make you overlook the fact that there is a second frontier. This I discovered when an undergraduate at Cambridge. Each year I used to design and, in the vacations, make and race cars of unorthodox design and eccentric behaviour.

On this occasion I built up a special engine based on an Austin 7 to go into a very lightweight chassis. A redesigned cylinder head and induction manifold combined with two huge carburettors boosted the power output to an astonishing degree during the short period before the crankshaft became tired of it all and disintegrated. To save weight the chassis was drilled for lightness everywhere and, as the first event for which it was entered was a sprint hill-climb, more weight was saved

8

by replacing the normal radiator with a zig-zag aluminium tube.

The car was not completed until the early hours of the day of the event and I duly found myself on the starting line with a car that I had never driven, facing a hill I had never driven up. Halfway up the hill there was a long left-hand bend where a friend of mine was stationed to take photographs with my camera. I hoped to learn something about suspension deflections and roll from the pictures.

I duly left the starting line with the back wheels spinning furiously in both bottom and second gears. We reached the long left-hand bend and again I opened the throttle fully. The rear wheels spun wildly, flew sideways, and the car started to spin like a top. As there appeared to be nothing positive that I could do about it, and as we were heading for a very substantial-looking barrier, I hurriedly left over the back of the car and watched its subsequent career with interest as I slid along the road.

Immediately the car appeared in sight, the more intelligent of the spectators had taken to their heels, and by this time all of them were in full flight. All except one: my friend, showing courage over and beyond the call of duty, was standing firm and took a photograph I still cherish. A fraction of a second afterwards the car ducked under the barrier and ran over him. I felt gratified that I had reduced the car's weight to a minimum. The camera departed like a well-hit cricket ball.

After a few days in hospital my friend completely recovered but unfortunately the camera was never the same again.

The moral of this deplorable incident is that you must

not overlook the possibility of a second frontier. If one lies through the cylinder head there will be a second one where the tyres touch the road. The performance is not only limited by the power that the engine can transmit but also by the amount the tyres can transmit. At the first frontier the gases cross from the inert to the energetic. This energy is then imprisoned in the transmission, apart from a small fraction departing in heat due to friction or turbulence, until it arrives at the decisive frontier where it tries to push the car forwards by pushing the ground backwards. Here it can escape at last and dissipate much of itself in heat by slipping if it is allowed to do so.

To create an efficient car, start with a cylinder head and four tyres and design in and out from these frontiers. The successful Rover 2000 was possibly the first car consciously designed in this way.

The next step in considering these starting-line frontiers is to realise that although they are often straight lines there is no reason why they should always be so. An electric motor is a self-evident example. Here the frontier is curved, lying down the centre of the air-gap between the rotor and stator. What is designed each side of this energy-transfer area is decisive. Of course if you take the curve and bend it into a straight line you find you have a linear motor (and don't forget that you can have a linear d.c. motor by doing the same thing).

Now by this time we are all becoming a little tired of the word 'frontier' as it is an imported geographical term and has no scientific significance. Cannot we find something less vague but equally general? To deduce a general definition from only three particular examples is asking for trouble, but at least they help to eliminate a host of possibilities as they already have so little in common. In fact the only shared characteristic is that their frontiers are all at energy interfaces. In short, unless we say in general that we must select our design point at an energy interface, we cannot say anything in general at all. It is that or nothing. We must now consider if there is, in fact, much likelihood of this definition being more than a local peculiarity, and we will do this by applying it to other design situations and see where it leads us.

One's first reaction may be to say that in problems of statics it is not going to lead us anywhere as an energy interface is a dynamic conception. Anyway we will have a look and see.

A bridge designer is always faced with two embarrassing possibilities. His bridge may feel as if it is going to fall down but doesn't. Alternatively, it may feel as if it is going to fall down and does. The unsafe feeling is caused by the bridge deflecting to an alarming degree.

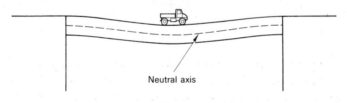

Neutral axis

Fig. 2

Fig. 2 shows a lorry passing over a mild steel bridge made in a highly simplified design. The bridge supports the lorry by failing (slightly) to do so. It deflects enough to gather a compensating reaction. The energy dissipated by the lorry losing height is absorbed in the bridge, much of it in the horizontal girder, where it is stored in

11

compression or tension movements in top or bottom parts of it. Between these energy stores there is an interface which lies along the 'neutral axis' whose position is a vital starting point in statics for designing the strength of the bridge.

Next we should look at what happens when a lorry, too heavy for the bridge, starts to cross it. As we have already seen at the beginning the interface lies along the neutral axis and the energy is stored in elastic deformations.

Presently the loads become higher than the elastic limit of the steel which suddenly yields and forms a series of hinges, as in Fig. 3.

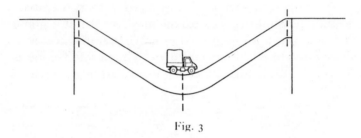

Fig. 3

These hinges now dominate the situation as the steel collapses in and out from the vertical interfaces. As this failure is sometimes multi-dimensional it might be better to say a point on the interface. Thus for calculating deflections we use the interface corresponding to the neutral axis and for failing loads the vertical ones at the hinges.

And the same criterion applies if we lengthen our bridge until stability becomes the chief problem. Torsional vibrations along it, rather than traffic over it, must be protected against. Our energy interfaces become an infinite number of discs threaded on to the longitudinal axis of the bridge. The more the discs become like rings, the more efficiently will they transmit the energy for a given mass, and so you end up with a tube. For convenience of construction and use, you squash this into a rectangle and so evolve the box-section girder bridge. Provided this doesn't fall down when you are putting it up you have an economical structure, designed for torsional strength, which has as a fringe benefit the ability to carry the vertical loads of the traffic over it. Each piece of steel can do two jobs at once.

If this example gives us an uneasy suspicion that there is more in statics than statics the next one will confirm it. It arises from the latest research into machine tool design. In the past the general concept of the design of a frame for a machine that removed metal was that it should hold the cutter or tool reasonably firmly against the object. Any small deflection could be compensated by moving the cutter up a bit. We now know how misleading this idea can be, for the reaction between the cutting tool and the metal is not wholly a static one (see *Machine-Tool Dynamics* by D. B. Welbourn and J. D. Smith, C.U.P., 1970). Irregularities in the metal result in a corresponding varying reaction and this energy is transmitted out from the interface into the framework on each side. And this framework is, itself, in the words of dynamic control theory a 'closed loop'. Thus to design a chatter-free machine we must start at the cutting interface and consider the hazards of resonance in the framework on each side and what can be done to avoid them. And as we go on our calculations follow the pattern of orthodox feedback and control technology.

We have to solve statics by dynamics, starting at the energy interface and facing the implications of its presence.

Summarising, this chapter is not concerned with proving anything but with suggesting a possible way of escaping from a dilemma. If you are puzzled how to select a starting point for designing, it suggests that you try working outwards from the energy interfaces.

2

SELECTING THE POSSIBILITIES

Having selected the focal starting point, or points, we must now turn our attention to the actual design procedure itself.

Many excellent systems of design techniques have already been evolved, aimed at achieving the ideal solution in a series of logical steps, and these help many people. Others find them valuable but for another reason. For them, such a system is of little use in actually creating anything new as they do this by their own untrammelled inventive flair, but, having settled on their design, they subsequently find some of these techniques most useful in proving how logical it all really was. Serving it up in that way often makes a better impression on bankers and accountants.

In view of the excellent methods already available I shall not attempt to provide yet another; my aim is a humbler one. I want to give a little help to someone whose mind is more or less a blank. An aid to get the party going, while, at the same time, leaving all the doors open for unexpected but welcome guests in the form of sudden creative inspirations.

As we have already discussed, your strategy will be to concentrate on a selected frontier, as the impact of the whole project and its multitudinous requirements is too intimidating. My suggestion is that, to de-blank your mind, you should simplify matters still further and try solving this primary problem by listing the ways in which it could be done if one of the design requirements was omitted. Then go back through this list and find which could most easily be adapted so as to re-instate the missing condition. The same technique can also be used at any point in the design and is not only confined to interfaces.

Practical examples may be the best way of explaining this technique, and we will start with a comparatively unsophisticated one.

Reverting to Fig. 1, it will be seen that the press descends into a tray of material, and, to ensure that the edges are properly formed, it is essential that there are no escape passages around the periphery. The clearance between the edges of the press plattern and the walls of the tray must be a few millitmetres at the most. The tray itself, being some 1.2 m wide and 5 m long, is very heavy and must be moved into the correct position with speed and stopped with precision. If it is a little out of place the press will come down with one side on the wall of the tray and produce instant and expensive chaos.

So expensive that everyone is threatened with extinction if it is ever allowed to happen again. As a result there is overwhelming public support for the idea that a consulting engineer (who happened to be myself) should be called in and made totally responsible. So I go and look at the machine. The first thing I see is that the tray, or rather the trays, for there is a long line of them, are linked by a conveyor chain which provides both the moving and braking forces. Conveyor chains have long links (for they are cheaper than short ones) and so allow wide teeth on their sprockets (which are stronger and less vulnerable to wear). But they have one characteristic which may prove embarrassing. A uniform

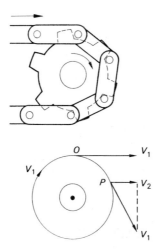

Fig. 4. It can be seen that at O the horizontal velocity of the chain, V_1, is the same as the circumferential velocity of the chain wheel. At any point further round, say at P, the horizontal velocity of the chain will be V_2, i.e. the horizontal component of V_1 at P. The distance OP depends on the pitch of the chain, and after it exceeds half the length of a link the next tooth begins to drive and V_2 begins to approach V_1 again.

angular rotation of the sprocket does not produce a uniform linear movement in the chain. As shown in Fig. 4, where the pitch is exaggerated for clarity, the chain has a wave-form superimposed on its mean velocity due to the wide spacing of the sprocket teeth.

Now, in the circumstances, this variation is highly undesirable, but to replace the chain with a fine-linked one will be a major modification. Can we modify the drive to the sprocket so that the chain will move, to all practical purposes, in a linear manner? We can split this problem into a basic one and a subsidiary one. Basically

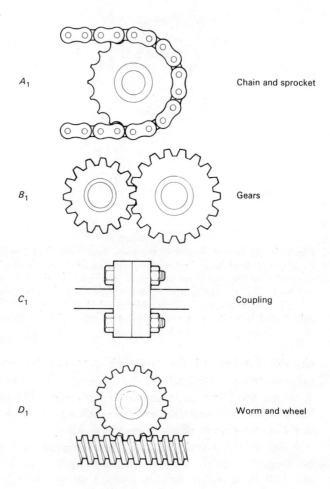

A_1 Chain and sprocket

B_1 Gears

C_1 Coupling

D_1 Worm and wheel

Fig. 5. Basic methods of rotating a shaft

we must drive the sprocket round, and secondly we must so arrange the drive that it produces a constant velocity in the chain. If we cannot immediately think how all this can be done the technique is to concentrate first on the basic condition only and forget the other. What are the possible ways of rotating the shaft on to which the sprocket is keyed? Let us start off with that and draw the various possibilities on the assumption that chain velocities are irrelevant. Fig. 5 shows a number of them.

The next step is to re-impose the second condition and modify each method of drive in turn to comply with it. Fig. 6 illustrates how this might be done. Some of the mechanisms are obviously clumsy but we are not evaluating them at the moment, this comes later, we are only concerned here with exploring possibilities. At least our minds are no longer blank. But there is another, and possibly unexpected, bonus that may arise from this method. While we have been studying the problem, our subconscious mind may well have been working at it too and may suddenly hand up some quite different solution, and possibly a much better one. It may suggest that we control directly the movement of the particular tray in the press area and only use the chain to tow the other trays after it. Or else it may suddenly occur to us that if the distance from the centre line of the press to the axis of the driving shaft was exactly divisible by the pitch length of the chain, any inaccuracies would always exactly recur and so could be compensated for by some fixed delay or distance.

Another possibility might be to abandon any attempt at adequate metering through the chain complex at all, but to provide instead proximity detectors that will

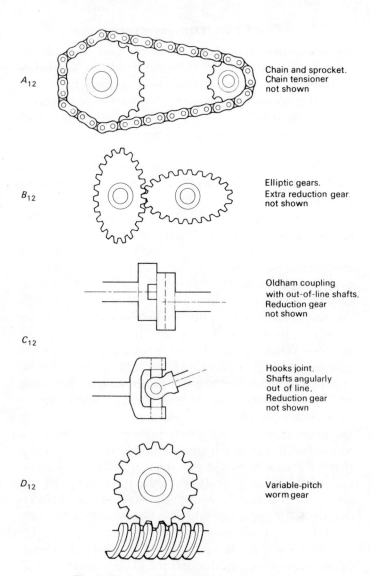

A_{12} — Chain and sprocket. Chain tensioner not shown

B_{12} — Elliptic gears. Extra reduction gear not shown

C_{12} — Oldham coupling with out-of-line shafts. Reduction gear not shown

Hooks joint. Shafts angularly out of line. Reduction gear not shown

D_{12} — Variable-pitch worm gear

Fig. 6. Basic condition plus second condition

control the power, positive or negative, to produce an exact register directly relative to the press itself. And so on. By this time our mind may have passed from being too empty to being too full, and in later chapters we will try and deal with this problem, but meanwhile we have, at least, got the party going.

This method is not limited to two design conditions only, it will work with three or more just as well. For instance, it can be applied to the problem of designing the front suspension for a car, a multi-condition complex. Here again we start with the basic requirement, and this is that something flexible must be placed between the body and the axle. Later we will feed in the other conditions in turn, preferably in order of importance.

Fig. 7 shows a front axle in its simplest form at A, and above it, shown dotted, the lower part of the body or chassis of the car. We must now list all the ways we can think of, without becoming mechanically ridiculous, that will directly support the body and absorb bumps. To save space we will only draw one half of the symmetrical layouts.

B_1 shows a simple coil spring inserted between them.

C_1 widens the spring base by extending out sideways the mounting on to the chassis.

D_1 mounts the spring at an angle to avoid having an extended mounting.

E_1 shows a semi-elliptic leaf spring extending longitudinally beneath the chassis frame member. It cannot be mounted further out without limiting the steering lock.

F_1 is a quarter-elliptic leaf spring mounted transversely.

G_1 shows a single half-elliptic transverse leaf spring

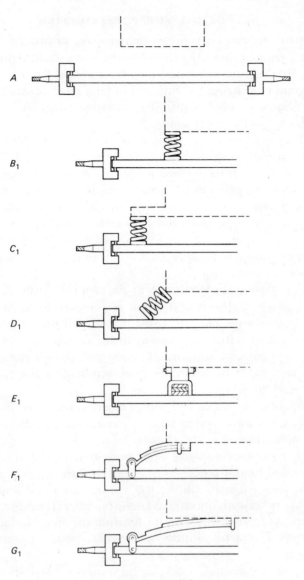

Fig. 7. Basic condition

attached to each end of the axle with a central mounting on the body.

Theoretically we should also have included tension springs but, although they were used in the early days of motoring, no insurance company would accept them now. A broken compression spring will continue to support a car, after a fashion, but a tension spring will not. In any case a tension spring design is the same as a compression one if you imagine gravity acting the other way up.

Torsion bar springs are not included as a basic possibility as they cannot be directly attached unless you twist them into a spiral and so end up with a helical one.

Suspension systems of this general type were suitable in the early days of motoring but as the speeds of cars increased people began to realise that springs were needed not only to make the car comfortable but also to make it controllable.

A wheel cannot receive any reaction from the road if it does not happen to be touching it. And without a suspension system a car would fly through the air with the greatest of ease.

Imagine going over a hump-backed bridge in a car with no springs. The car, and its wheels, could only accelerate downwards at $1\ g$ and so might easily jump a long way. If springs are fitted they will exert a total downward force equal to the mass of the car. If the unsprung mass, i.e. that of the wheels and axles alone, is a tenth of the car mass, the springs will (anyway for a short distance) produce in them a $10\ g$ acceleration downwards. For simplicity we are ignoring the damping of the shock absorbers. Springs stop a car behaving like a tennis ball.

23

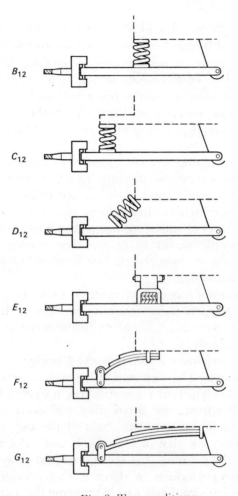

Fig. 8. Two conditions

24

It follows that the more we reduce the unsprung mass the more stable our car will be. This condition we will now superimpose on our basic designs. As in Fig. 8, B_1 becomes B_{12} as we halve the unsprung part of the system by cutting the axle into two halves, both pivoted in the middle of the car.

C_1 similarly becomes C_{12} and D_1, D_{12}. E_1 follows suit but its leaf spring will now have to withstand some torsional stresses as well as bending ones and may not like doing so. However, F_1 and G_1 will accommodate the divided axle layout quite happily, as in F_{12} and G_{12}.

The next stage in the evolution of car suspension systems was that of absorbing the conditions arising from the use of wider section tyres. The wider the tyre the greater the friction, always providing that the wheel remains substantially vertical to the road surface. Otherwise only a corner of the tyre would touch and the total situation would become worse rather than better.

Thus, to exploit these better tyres we must have them moving up and down vertically when hitting bumps. This, then, is our third condition, and is superimposed on the first two in Fig. 9.

B_{123} has two equal-length arms which make the wheel move parallel to the body of the car. C_{123} allows the steering assembly to move up and down a tube within the spring, and by letting it rotate around the tube too, we can dispense with the separate king-pin and save more unsprung mass.

D_{123} puts the compression spring along a diagonal of the parallelogram formed by the arms.

The evolution of E has been that of changing in steps from a bending to a torsion moment in the spring. E_1 has very little torsion, E_{12} an appreciable amount. In

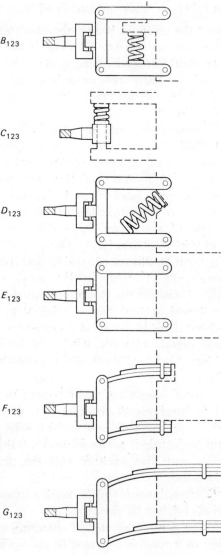

Fig. 9. Three conditions

26

E_{123} the lower of the parallel arms is pivoted immediately above the centre line of the longitudinal spring, which will now need to withstand torsion only. To cope with this effectively we must turn our leaf spring into a circular bar, one end of which rotates with the lower arm and the other end of which is fastened rigidly to the chassis. F_{123} and G_{123} duplicate their systems in parallel.

Now that we have achieved, in stages, a much greater reaction force between the wheel and the road we find that there is an awkward side-effect. Making the wheels move substantially parallel to the vertical line of the car is ideal for bumps and reducing tyre wear, but it also permits much greater cornering forces. This results in the body of the car rolling out of the vertical and the wheels following suit.

Our fourth condition is therefore that of finding a means of geometrically adjusting our systems so that the extra cornering load on an outside wheel, and its resultant suspension deflection, may be exploited to restore the wheel to the vertical. Of course this cannot be done exactly and there are other factors involved, but we must have a means of modifying the strict parallelism that we have at the moment (see Fig. 10).

Systems B, D and E can be adjusted by making the lower arm a different length from the upper one; generally it needs to be longer. F and G will have to be abandoned; they are too unpredictable and sideways forces upset their symmetry.

L_{1234} only works because there is some built-in flexibility in the top mounting of the tubular strut.

In practice all the designs shown can use ball-joints to provide rotation in two directions at once, so eliminating separate king-pins, and in fact this is nearly

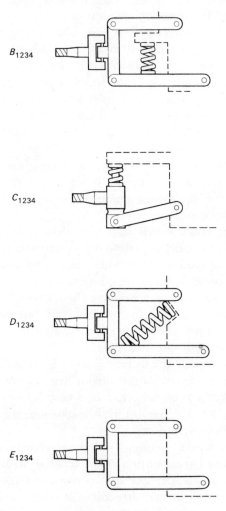

Fig. 10. Four conditions

always done. However, to preserve diagrammatic clarity, we have not incorporated this refinement in the drawing.

Summarising our procedure we can say that, without any great inventive paroxysm but merely by following our noses, we have systematically evolved four basic suspension systems, B_{1234}, C_{1234}, D_{1234} and E_{1234}, each complying with our four conditions. And in fact these four cover over 96 per cent of the designs actually used in the cars currently available on the British market.

Here again, we must leave the consideration of our final choice to a later chapter.

So far we have illustrated this technique of piling up design conditions, by applying it to simple and complex versions in turn. That the examples have been taken from energy interfaces already discussed is to avoid proliferation of problems and to preserve continuity with the previous chapter. There is no reason why this approach should be applied solely to these critical points; you could use it anywhere if you felt it might help.

Finally, it is important to remember that when you design a machine people will expect it to work. Alibi-hunting afterwards will do little good to anyone. If a designer is responsible for a machine it is no good him stating afterwards that he was given wrong information. Part of his job is to query any conditions about which he has the slightest suspicions, even if they are outside his area of expertise.

Problems of this kind sometimes arise when firms diversify into new fields outside their direct experience, especially when this involves some new chemically based condition being added to the mechanical ones. Although not a chemist himself, the engineer must do his best to vet the new requirements.

29

I was once asked by the Board of a company to design a production line for a new product based on the application of a novel chemical technique. I said I would like to know rather more about this side of it, if the chemists could find time to instruct such a novice in the subject. In due time a visit was arranged and I arrived at the factory. The managing director received me warmly, gave me coffee, explained that all my questions would be answered by the technical director and passed me on to him. He said that all the information was collected in the research department under the chief chemist, and sent me round in a car. The chief chemist said that there was a little confusion about the matter and, in fact, it was the production chemical department that had done all the work. Another car journey, for it was a big organisation, and I was being courteously received by the works chemist. He said that, although he had signed all the reports, the actual work was done by one of his assistants, quite a young man, and sent for him. In the interval I asked what the young assistant's chemical qualifications were, and to my surprise they seemed to overlap very little on to the new venture.

However, everyone regarded him as an expert and I awaited his arrival with interest. He entered, obviously an intelligent man with probably a good future, but what first caught my eye was a black cylindrical leather-covered box he was carrying. We all sat down and the conference started. The young chemist opened his box and produced an ancient and large cylindrical slide rule. Each time he was asked a question he would twiddle this slide rule. It seemed to mesmerise my companions. Ask him a question and there was always this pause for a fiddle, which gave him plenty of time to think up an answer, which, when it came, seemed in the circumstances to carry an almost pontifical authority. I had a nearly unbearable desire to ask him what the time was; I was sure he would consult his tabulated cylinder first.

I thought it wiser to advise the Board that more objective chemical data should be available before any more engineering work was carried out.

3

SELECTION BY CONTEXT

The first step is to consider the particular context within which the design will have to work. As usual, we will introduce the idea by as elementary an example as possible. You are asked to select a two-way air valve to perform a particular function and you say that you cannot do so until you know where it is to go. If it is to go into a steel mill your design must be quite different from one needed, for instance, to operate a dentist's chair. In a steel mill no one ever turns a valve on; they hit it. They snatch up the heaviest blunt instrument, generally a spanner or crow-bar, and you may well have to choose the valve with these knock-out blows as a major design consideration. For doing the identical job in a dentist's chair you need only have a little valve with a knob suitably polished to avoid scratching the finger-nails of his glamorous assistant. The context is decisive.

Now although concrete working conditions are part of the context they are not necessarily the whole of it. Abstract considerations such as cost, reliability, accessibility, appearance, and so on, may well have something to say about which design we select.

In short, we construct a framework of relevant considerations and assess each design by how neatly it fits in.

Of such containing contexts there are two particular ingredients that are only too easily forgotten. Not, paradoxically, because they are abstract and unusual, but because they are concrete and commonplace.

To cite topical illustrations is tempting, but a consulting engineer who tells tales about his clients will soon have no clients to tell tales about. So I must go into the past, I hope the forgotten past, for my illustrations.

I remember quite vividly the first production run of a new machine which, for its day, was, I hoped, a pioneer in its field. Its control system was a pneumatic one and quite involved. The machine had been tried out the night before, and today it was fully crewed up for the first continuous production run. Various worthy gentlemen, who had financed it all, were reputed to be coming along to see it in action. Everything went smoothly for nearly an hour, and then the main control system went mad. The machine appeared to be having a smash and grab raid with itself, and the cause was only too apparent. I had forgotten something obvious: I had forgotten that the factory was full of air. Air is full of water vapour, and water freezes when cooled by a powerful jet of air escaping from a largish valve. In short, one of the main valves had coated itself in ice and packed up. Everything and everyone stopped and looked at me. I hurriedly took some pound notes out of my pocket, gave them to the nearest electrician and told him to hurry to the neighbouring shopping centre and buy a ladies' hair dryer. Smirking happily, he rapidly departed.

All the same my mind was not entirely at ease. I wondered what the worthy gentlemen would think as they inspected a machine of such reputed sophistication and saw, in the centre of it all, a ladies' hair dryer coloured, as it turned out, in a particularly revolting pink.

By next day properly designed modifications were in place and the pink eyesore banished, but it had been a tricky few hours. Air is nearly always part of a machine's context, so remember to consider whether its invisible and often corroding contents are a relevant factor.

The other common context, which is commonly forgotten, especially by young designers, is the diabolical effects of vibration. The fact that you have shown a bolt with a nut on it on a drawing does not automatically mean that the two will remain together very long when the machine starts up. I doubt whether this problem is ever sufficiently appreciated until you have seen some spectacular disaster provoked by it.

With me, it was the experience of a friend of mine in the early days of motor racing at the old track at Brooklands. He was determined to break the hour record for three-wheeled cars. Someone had travelled over 105 miles in an hour and he was determined to better it. He had a Morgan three-wheeler fitted with the J.A.P. engine mentioned in the first chapter, which propelled it along at over 110 m.p.h.

At first he found the direct steering of the Morgan very difficult, for it made the car behave like an apprehensive spider, but he improved matters by bolting in a little reduction gear, off a model T Ford, immediately under the steering wheel. He also used a hand throttle, off a motor bicycle, so he could set the engine flat out and leave it there.

He then started his record attempt, circling the pear-shaped track where he periodically disappeared behind a hill out of view of his friends at the pits. After a few laps, and when out of sight, all the bolts shed their nuts and the reduction gear came in half, and my friend was left with a steering wheel not connected to anything but one little gear wheel. He tried in vain to re-insert this back into the rapidly disintegrating gear assembly, gave it up as a bad job, and jumped over the back of the Morgan. As the hand throttle was still in the all-out position the Morgan, relieved of his weight, re-appeared out from behind the hill going great guns. It safely negotiated the banked curve and then set off down the straight, but at a slight angle to it. After a few hundred yards it deserted the track, knocked down some railings and plunged enthusiastically into the sewage farm.

His horrified friends raced after it and searched wildly for my friend around the partially submerged Morgan. They were both relieved and annoyed when he appeared in view, quite unhurt, walking down the track. Remember that most machines have moving parts and this means vibration, and persuading a nut to stop on a bolt is a science all of its own. No designer should neglect to study it, or at least know where to find the advice of those who have.

But we must not forget that vibrations are produced by things that go up and down as well as those that go round and round.

For instance, imagine that you are designing some heavy earth-moving machinery with hydraulically operated scoops, rams and so on. You ask specialised hydraulic manufacturers to quote for a suitable system. One uses oil at 1000 lb./in.^{2}, another quotes for 5000 lb./in.^{2}. Which do you want? The cheapest, please. That means the higher pressure one, for the cylinders and so on can be much smaller in diameter. But it will not be the cheapest when you relate it to the context, i.e. the up-and-down crashes and vibrations invariably produced as the machine plunges and bangs its way round a building site.

The poor structural rigidity of the smaller diameter

cylinders and piston rods will make them too easily vulnerable to fatigue or accidental damage and so they will need separate support. It will be cheaper overall to select the lower pressure, sturdier installation.

And there are subtler dangers too. Design a self-supporting structure or component with a uniform factor of safety to bring down the cost and you may well end up with what you aimed at, i.e. a constant degree of rigidity throughout. Now put this component in the context of high frequency vibrations – and frequencies are getting higher all the time as shaft speeds increase – and you may find that it starts to resonate at its natural frequency. And as you have designed it so uniformly the whole thing will resonate at once, instead of each bit in turn, and the forces will be alarming.

Next we must consider more abstract requirements. In Chapter 1 we evolved a number of ways in which we might precisely move and locate a tray relative to a press platten. And the overwhelming context was that of reliability. Which system should we select? Now there are two different limiting possibilities here. Must the machine always continue to operate or can we let it stop in an emergency? In this instance the production line can stop, at any rate for a short time, without running up excessive expense. And the principle to adopt is not that of selecting one system but of selecting as many as possible that can be simultaneously fitted in. And if they start contradicting each other arrange for everything to stop instantly. The last three suggestions on page 19 would do. The point is to select systems so different in principle that anything that is likely to upset one is irrelevant to another. Segregate them completely, and if then they all agree it should be safe to carry on.

37

But what should you do if you are designing an aeroplane or a space-ship, where you cannot hang about while mechanics do repairs? In this context you must employ a different strategy. You must decide which system is most likely to be reliable and duplicate it, or treble it, keeping each system quite separate. Arrange things so that if any one system breaks down the others can carry on unimpeded without it.

In short, you need to have twins (or triplets) but whether to choose identical ones or opposite ones depends on their environment.

Next we will revert to our basic problem of how to achieve a uniform conveyor chain velocity, but move it sideways into a steel mill or down into a coal mine. In these instances a steady speed of the chain may be required to safeguard against longitudinal vibrations building up in the chain. Whether the system is to convey iron ore or coal, the context demands the same characteristics, i.e. low capital cost, ruggedness, reliability and readily available spare parts. This context will bring us back to consider the mechanical arrangements in Fig. 5.

The chain, and gear and worm solutions are awkward and expensive, so we are left with the ,two couplings: the Oldham and the Hooks joint. Of these the latter may be better from the point of view of keeping the lubrication in and the dust out, even though a pad type may have to be used. Again context is decisive.

Now we must consider how to deal with a context containing a number of factors. The suspension systems of Fig. 10 are an example of this. Taken in isolation each fulfils, in principle, the internal mechanical conditions. To make a realistic choice we must put them into the appropriate external and largely abstract contexts.

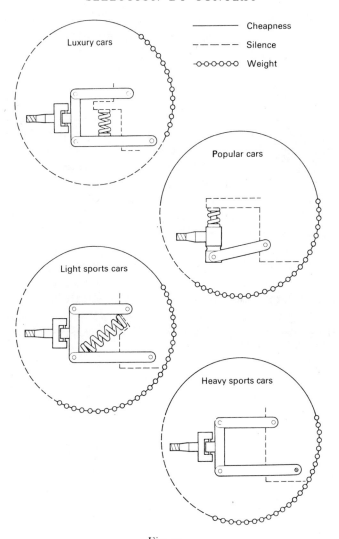

Luxury cars

Cheapness
Silence
Weight

Popular cars

Light sports cars

Heavy sports cars

Fig. 11

Amongst these are manufacturing cost, noise transmitted from road, and weight. From the cost angle C_{1234} is probably the best. It has fewest components and the attachment points to the body are sufficiently widely spaced to allow direct mounting on the body and so make a subframe unnecessary. D_{1234} is quite good as the spring and top arm share a common mounting. From the noise point of view C_{1234} may well be the worst as road vibrations would be transmitted straight on to the body and a subframe would be too spreadeagled. For sound insulation a subframe resiliently insulated from the body is ideal and can be done more easily with B_{1234} and D_{1234}. The end of the torsion bar in E_{1234} can transfer some noise, which makes it not quite as good.

For keeping down the total weight involved D_{1234} is possibly the best, unless the car is large and heavy, for then the diagonal spring would have to be unnecessarily heavy and its horizontal component across the car might make a bulkhead essential. If you want to keep the frontal area of the car as low as possible, select D_{1234}, or for a heavier car, E_{1234}.

We might summarise our selections according to the various contexts by giving each characteristic a length in the circumference of a circle indicating its relative importance, and putting the most appropriate design in the centre, as in Fig. 11.

4

SELECTION BY CONTENT

We will now imagine that we have before us two designs (or more if you wish), which seem equally preferable from all the considerations of context. We must therefore overcome the dead heat by assessing the relative values of their mechanical contents. And I suggest that we might systematically do this by comparing them in the light of some general principles.

The first of these is that a determinate design is preferable to an indeterminate one.

All mechanisms are excellent mathematicians. They scrupulously behave according to the principles of their construction. And the more accurate our mathematics about them is worked out, the better we shall be able to forecast how they will work when made. If the calculations are indeterminate the forecast will be indefinite. We shall only know exactly what will happen when the machine is made, and it may then be too late. The evolution of aeroplane design illustrates this. In early designs things fell off. To stop this the next generation of planes used two or more members to do any one job, feeling there was safety in numbers. This meant a host of redundant situations occurred. As creep and fatigue are not respecters of redundances things still fell off, the only difference being that you could no longer work out in advance the order in which they did it. Both the maths and the future were indeterminate.

But to do the calculations properly would take an

impossibly long time, especially when there were more than four variables, and so various rules of thumb were adopted. With these you could design a plane fairly easily. It was much more difficult to summon up enough nerve to make or fly it. It was only when the computer arrived that the modern giant plane could do so too. We could design determinately. In short, when choosing between two designs of otherwise equal merit select the one you can analyse best. The more you can calculate the less the gamble. Beware of the unpredictable menace of indeterminacy.

Now if both your designs are equally determinate you must seek a further criterion of comparison and, I think, it should be this.

'If one design is simple and the other complicated, choose the complicated one.' We must recognise that over the whole field of engineering endeavour an advance into efficiency generally means a retreat from simplicity. Engineers have been often mesmerised by the idea of simplicity, as if there was some inherent virtue in it. You selected, they felt, a simple design because it would be cheap or reliable. But what you were doing was to select the design, not, in reality, because it was simple, but because it was cheap or reliable. You told yourself that you were selecting the design because of its content; actually you were doing so because of its context. Worse still, the pursuit of the fictitious virtue of simplicity sends engineers bolting down culs-de-sac.

In the early days of motoring the two-stroke engine was greeted with shouts of joy, it was so simple. And it was a failure. The four-stroke, side-valve engine, much more complicated, became the fashion, and did well. Next, searching after greater efficiency, the push-rod overhead valve was evolved and today there are virtually no side-valve engines made. And now the push-rod engine is being replaced by the overhead camshaft engine with a chain or notched belt drive: more and more complexity providing more and more efficiency.

Now if there is no intrinsic virtue in simplicity, is there any in complexity?

Complexity can have two distinct meanings in design. First there is the complexity of untidiness. One way to design a machine is to divide up the problem into sections, solve each in turn, and put your solutions end to end. But if you divide up you must always tidy up afterwards. Complication has no virtue in it, if it is merely another word for a mess.

The second meaning it may have is that of precision. The fundamental fault of the two-stroke engine for motor cars was that its simplicity really meant compromise.

In general, any basic principle becomes increasingly efficient with the precision of its application. The evolution of combustion chamber design was due to this compulsion of precision with complexity only as a characteristic by-product. Figs. 12 to 15 show the sequence whereby the b.h.p. per litre was multiplied four times as the simplicity of compromise was supplanted by the precision of the critical.

The basic two-stroke design (Fig. 12) has so few parts that each has to do several jobs at once. The piston has to act as a couple of valves as well, which means that a quarter of its movement is ineffective, and it has an unsymmetrical deflector, with a high mechanical nuisance value, perched on its head. The crank case has to be a compression vessel as well. Jacks-of-all-trades

43

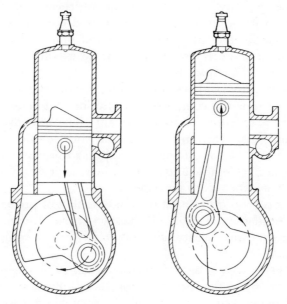

Fig. 12

abound. Nothing can be precisely designed because nothing is precisely done.

Fig. 13 is the simplest form of the four-stroke engine, where we can separate the jobs up better. The valve timing can be quite precisely arranged but the combustion chamber is a compromise. Its shape is determined as much by the necessity to fit the valves in as by its effectiveness as a piston pusher. To increase its efficiency you must raise the compression ratio; but if you do the valves cannot open so far, and down the efficiency goes again. Compromise is still rampant.

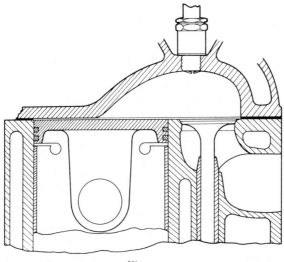

Fig. 13

Matters are improved slightly by putting one of the valves on top, as shown in Fig. 14, as both can be made larger, but we are faced with more complication if we do it. In Fig. 13 each valve could be directly pushed up and down by a camshaft conveniently located for driving off the crankshaft. Now the overhead valve will need a longer rod and a rocker arm, but it will be worth it, as we also have the extra bonus of easier gas flow.

The next step would therefore seem to be that of putting both valves in the head, as in Fig. 15.

Now, at last, we can have a little elbow room for designing our combustion chamber shape, but it is not very much. We still have to have a flat top to it for the valve seats. In addition the increasing inertia of the

45

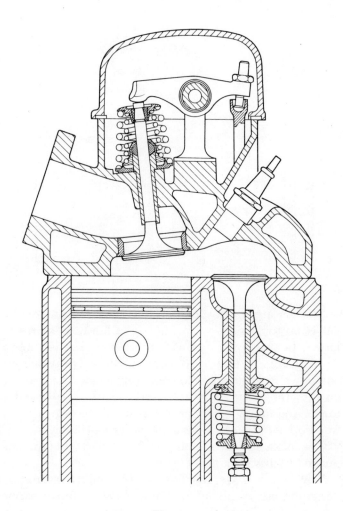

Fig. 14

46

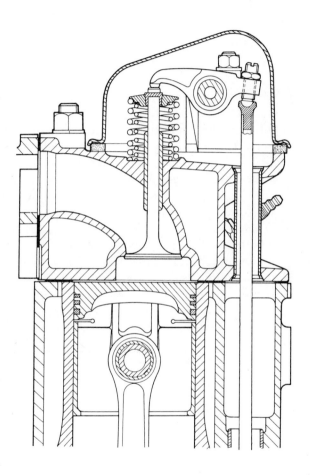

Fig. 15

valve operating gear is now beginning to limit our maximum r.p.m. as valve bounce will set in and we shall lose our timing precision.

There is little that we can do unless we take the complicated step of lifting the camshaft bodily up on to the top of the cylinder head, which means a much more involved drive, but we will be able to control our valves better.

This precision-based complexity is equally relevant to the design of process machinery.

I remember the first continuous-production machine of importance that I had to design. I worked on two different principles and had to choose between them. For those days the designs were quite sophisticated; nothing quite like them had been attempted before anywhere in Europe. Both needed involved control systems and it was here that the basic difference arose between the two. The first machine employed a precise mechanism to control each function. The second machine grouped the functions together and controlled them in pairs or more. Although the percentage capital cost between the two proposed machines was marginal, the second design looked infinitely more attractive; it was much simpler; I selected it and I thought I was being clever. I was not; the second machine was built and it worked, but there was always insufficient scope for dealing with eccentricities in the raw materials. I had been too simple-minded.

In short, this criterion of selection is the reciprocal of the last one, where we sought to avoid redundancies. Redundancy means too many parts chasing too few jobs; over-simplification means too many jobs chasing too few parts.

In general, choose precise complexity rather than blurred simplicity.

At this point one begins to wonder if this phenomenon of precision-based complexity springs from something more basic than mere expediency or experience. And I think it does.

You will notice that the design of the elementary two-stroke engine can be easily visualised in terms of two-dimension images. Next, as you build in efficiency, you progressively force out two-dimensional thinking. In fact you can only picture a four-valve spherical cylinder head in three dimensions. Reality, too, is obstinately three-dimensional. Move from two to three dimensions, with its increased degree of freedom to be complicated, and you move inevitably from a simple world to a more involved one, but you move nearer to reality.

Early bridge designers were two-dimensional thinkers. Their bridges were less inhibited, they danced about in three dimensions and fell down because of instability. Reality was more complicated than was thought.

And this is in turn, I think, based on an even wider foundation, for the physical sciences also warn us against being too simple-minded.

Einstein could not be as simple as Newton, he knew too much. And now physics tells us that any mental picture of reality is a hindrance to the understanding of its nature. We are too unsophisticated. And so if your designs are to harmonise precisely with reality watch out for complications and be suspicious if you do not get them. But you must master them, not let them dictate to you.

So far we have been considering the mechanical content of our designs, but it is not on this aspect alone that our selection should be based; equally vital is their financial content. Machines cost money; no one wants to lend you money unless you can show how your project will pay it all back, including the interest on the money, depreciation and so on. An engineer has not only to make machines but to make money too. An added difficulty is that he has to show in advance how much he is going to make. And so he must have the costs and savings assessed, probably by the accounting department. Here again, as with chemical specifications, it is surely the duty of the designer to keep an eye on what is going on. It is unfair to expect non-technical accountants to act in isolation, nor must it be assumed that an accounting system, ideal in other ways, is automatically accurate when it comes to assessing the value of new processes or the machines to do them.

I remember working at a design in a ground-floor office of a large company. Searching for inspiration I gazed out of the window into the gloom of a winter's afternoon. And I suddenly saw a curious sight. An old man was walking past the window pushing a handcart which was piled high with some of the firm's finished products. As the factory had fully mechanised transport I wondered what was going on, so I slipped out of the office and followed at a discreet distance. The handcart wound its way through the factory until, at last, it came to a rubbish dump in one corner. I had never seen this before as it was hidden amongst the fifty acres of buildings. The contents of the handcart were tipped on to the dump and the man departed to re-appear a short time later with another load. And other men were doing the same thing. Each load and much of the contents of the dump consisted of the company's finished products, all slightly damaged, and therefore unsaleable. Three or four lorries a day arrived to empty the dump, which also included miscellaneous rubble and waste.

On investigating further, I found that, mixed with the waste in lorries, many hundreds of tons of expensive material were taken out and dumped each year.

To re-use the material would only need a simple machine to break it down, when it could replace the normal raw material costing over a hundred pounds per ton. It seemed an obvious thing to do, so I asked the accounting department to work out accurately a yearly saving.

They said it would save nothing at all. This I found a little curious, and I asked how they had reached this conclusion. They replied that as there was no figure in their accounts for disappearing products, none could be disappearing, and so nothing could be saved by stopping it. I thought for a moment and then said that, as my suggestion would provide, almost free of charge, many hundreds of tons of raw material per year, it could surely show up as an asset on the input side of the accounts. This was regarded as heresy. It was an inviolable rule of accountancy, I was told, that all raw materials must be shown at their finished cost value. As this raw material was the company's finished product it must be charged up with all the processing costs included. This would mean that it was worth three times as much per ton as the raw material and its use would therefore show in the accounts as a heavy loss. Even the most ideal accounts are sometimes not entirely realistic, and it is unwise for the engineer to accept all the figures without query. If there is a possibility that the accountants know as little about engineering as you do about accounts, remember that a little co-operation will help both of you and also the company you are serving.

You should also bear in mind that, when choosing between two possibilities, you should preferably select the one that could not have been designed five years ago. If both could be as old as that, do not select either. If you do, your competitors will have produced something better before you have made yours. In heavy engineering you may have rather more than the five years to play with, but in lighter industries, especially electronics, it may well be much less. In engineering races there are no cash prizes for being second.

But, of course, the most literal meaning of selection by content may well be the selection of content, the selection of the right material for the job. We must discuss this together.

5

SELECTION OF CONTENT

I have done all the hard work so far: it is your turn to
do some now.

'What do you mean?'

I want you to invent something.

'By myself?'

More or less. I'll try to help you if I can.

'What am I supposed to invent?'

Why not a substitute for wood? It is still the most
popular building material: find an alternative, preferably
a better one.

'Do you mean that I have to sit down and invent some
revolutionary new plastic?'

No; that is exactly what you must not do. It would
be a waste of time. Leave jobs like that to the chemists
and physicists, they'll be better at it. An engineer selects
and arranges combinations of known materials; not
unknown ones. He is an opportunist; he exploits for the
benefit of the community the discoveries of science.

See if you can produce a composite material with the
good characteristics of hard wood but without the bad
ones.

'What's so good about wood?'

You could have answered that yourself. Wood is very
useful in that it is strong, light and you get a good
volume of it for your money, but bad in that it rots,
catches fire, gets eaten and warps.

'How am I supposed to start?'

You may remember that in designing a mechanism we began with the various basic designs and then modified them to suit other conditions fed in in turn. To design a composite material, use the same systematic pattern of thought. Begin with basic constructional materials and then feed in modifications.

'Would steel, aluminium, nylon, concrete, glass fibre and wire do?'

Yes, but include cement instead of concrete; concrete is a composite material already.

'Now do I make a list of their physical properties?'

Yes, but in such a form that you can compare them realistically with each other and also hard wood.

'I don't quite follow.'

Well, we have already agreed that wood gives you a good volume for your money. Work out the volume of each of these other materials that you could buy for some unit of cost, and compare them. Wood is strong too; if you had to suspend a given weight from a fixed height a wooden tension member would be a good bargain.

Work out the relative costs of this tension member in other materials and compare these too.

And then do it for compression loads, i.e. the cost of a column to support a given load.

'Do I just work it all out and put it down in tabular form?'

I think you can do better than that: and this is, I think, an important point. Always use a diagram rather than words or figures if you can.

Much of the creative part of an engineer's work is done by his subconscious mind which hands up ideas into his conscious mind, i.e. his imagination, and the subconscious can only communicate by using pictures;

54

it has no vocabulary. Symbols or numbers are beyond it, or beneath it. This means that an engineer's mind is more tuned to diagrams than arithmetic. See if you can devise one for this.

'Suppose I used a horizontal distance to represent how much volume you get for your money, and two vertical lines for the tensile and compressive strength for a given cost; would that do?'

It sounds all right, but I think I should join up the three lines with a fourth to make an enclosed figure; shapes are easier to remember than distances. Remember that wires and fibres have practically no compressive strength as they are structurally so bad at it.

'I've done this in Fig. 16; I took the market cost of the best quality in each case.'

Yes, you've done that quite well. What strikes you as important?

'I can see why wood is so popular.'

Yes, I agree. But I think you can learn more than this from the diagram. It highlights the essential difference between the mathematician and the engineer.

'But I thought the more mathematics the better for making a good design.'

Yes; but mathematics as a tool, not an attitude of mind.

A mathematician is constantly seeking general solutions to general problems. Having found one, often expressed in the symbolic terms of algebra, calculus, etc., he can then put in real values and produce real answers. He works from the general to the particular.

In engineering design there is never a general solution to any general problem.

'Why not?'

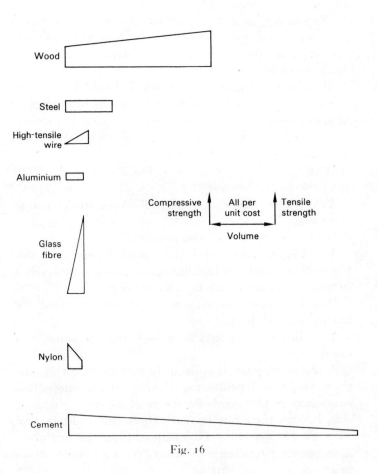

Fig. 16

Because there is no general material to embody it in. The design must be suited to the material and there are only particular, if not peculiar, materials available. In mathematics we go from the general to the particular, in engineering from the particular to the particular. Design a building in steel, wood or concrete, and your three drawings have little in common. Material and design are interlocked. You cannot design in a vacuum, only in a substance.

'Which comes first, selecting a material or a design?'

That's like the chicken and egg question. I suppose you can say of both that there is no general answer, only a particular one. But at the moment this question does not arise. Your particular problem has been decided for you: a substitute for wood.

'The diagrammatic shapes of the various materials tend to look like rectangles or triangles. Is there any significance in this?'

Yes; 'rectangular' ones are the more versatile; the 'triangular' ones are for special purposes and you may find it useful to remember materials in this way.

But there is a third, and very useful, significance in these shapes, in that you can combine them, and so make up new ones; new composite materials tuned into particular purposes. But in doing so you must remember one important fact.

When two different materials are stretched side-by-side one is almost certainly going to break before the other.

They will reach their breaking points at different extensions. Thus the maximum breaking force of the two together is not the sum of the two but that of the better of the two. And the same is true of compression

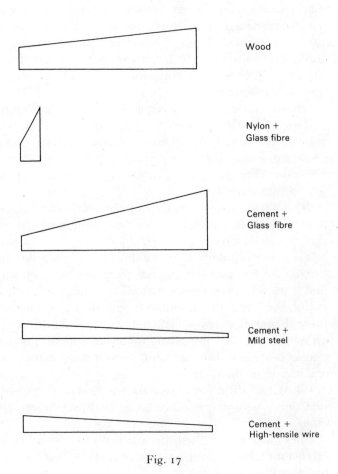

Fig. 17

58

as well. This means that in evaluating a composite shape the vertical components are the best of the individual ones. But you are supposed to be doing all the work this time; puzzle it out and see what you can come up with.

'Have a look at Fig. 17 and I'll try and explain what I've done. The total cost is the same as before but I've divided it between pairs of promising materials and spent half on each and combined them.'

Yes, that was the right thing to do at this stage. Now can you spot how to incorporate resistance against bending moments into the diagrams?

'Well, the tensile and compressive strengths obviously must still come into it and the volume has something to contribute too.

The more volume per unit cost, the deeper you can make a beam and so the better you can exploit your tensile and compressive stresses. I thought of saying that the area of shapes symbolised the bending strength, as this brings in all three factors. But it is not true of a purely "triangular" material which will have no bending strength at all.'

You are nearly there: have another think about it.

'Is it that the bending strength is represented by the area of the largest rectangle you can fit into the figure while still using the common base of volume?'

Yes. Do this and, for clarity, try shading in the bending strength area.

'Have a look at Fig. 18; cement and glass fibre are nearest to wood.'

Yes, but before we go any further, let's review what we have done so far. Without reliance on inventive thought, we have juggled about with market prices and arithmetic in a systematic manner and a composite

59

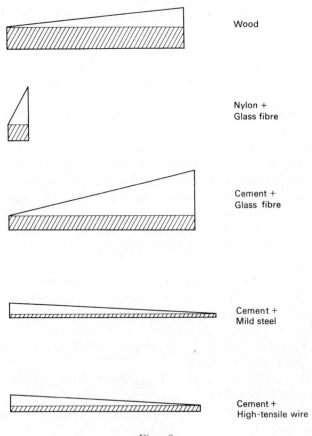

Wood

Nylon +
Glass fibre

Cement +
Glass fibre

Cement +
Mild steel

Cement +
High-tensile wire

Fig. 18

material has, more or less, selected itself. At the moment it is the method that I want you to remember. In any case, the conclusions may become out-of-date. Prices will have altered by the time this book is published, and anyway, other and better plastics and fibres will have arrived, but I think the method will still be valid for selecting potentially promising combinations of materials.

For the sake of our present study we will assume that the cement–fibre combination is our best bet.

Now we must move on from the realm of material selection of that of design technique.

'Do you mean casting or fabricating into special shapes?'

No; not at the moment. Obviously you can increase the effective strength of wood by sticking or screwing it together to form a special structure; similarly you can roll steel into sections, extrude or mould other materials, and so on. All this comes later. At the moment we are concerned with what can be done inside a material while leaving its outside still in the form of a rectangle.

'Where should I start?'

In and out from the middle, the energy interface.

'Well, I suppose the material in the middle does very little work for its living in bending-moment conditions: it just keeps the bottom and top faces away from each other. The glass fibre re-inforcements would be better employed if they were all concentrated near the surfaces.'

Yes, but you can do more than this. Air is one of the cheapest and most useful building materials and the supply position is good. Why not use some?

'You mean putting air channels down the middle?'

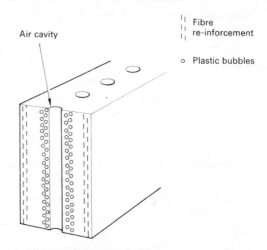

Fig. 19. Cut-through slab

Yes, but mind out you don't make them too big, so causing local weak spots. Remember a sphere is the best shape.

'Do you mean I have to blow a lot of bubbles and put them in?'

Yes, but you needn't blow them; just buy them. They are little polystyrene spheres full of air and a few milli-metres in diameter, very light and quite cheap. Now draw a section of what this imaginary material would look like.

'I've done that in Fig. 19, but it looks to me as if it will remain imaginary. I dare say you could assemble a bag of tricks like that if everything was dry, but trying to stick it all into sloppy wet cement and consolidate it properly would be an impossible job. While it remains dry it has no strength; and you can't wet it. Is a syste-matic approach to design supposed to lead you up the path into impossibilities like this? You've been wearing me out to no avail.'

Don't give up heart so easily, a good engineer wouldn't. Let's stand back a bit and look at systematic designing in general. It will lead you to one of three quite different positions. The first is that of having a variety of designs to choose from, and you can make your selection from considerations of context and content.

The second possibility is that you find there is only one design on offer that will match the conditions and so you must choose that anyway. Finally, as at the moment, you end up with an apparent impossibility, a self-contradictory design. When you start off you do not know in which of these three positions you will land, but if it is in an impossible one you are not necessarily defeated. I will try and explain what I mean and what you should do.

6

POSSIBLE IMPOSSIBILITIES

When you are faced with an apparently impossible situation, an in-built self-contradiction, I suggest that you try four different methods of resolving the problem, and you may well be able to think of others for yourself.

Now there is one strategy that you should firmly avoid, and unfortunately it looks superficially a highly attractive one. It is the first thing that would normally occur to you. You try to think out a compromise. Well, don't. Later on, when you have found the real solution, a certain element of compromise may be necessary over details. That is in order; but compromise over fundamental requirements is the escape of the faint-hearted, and you don't escape in the long run.

My first suggestion is that, instead, you should try to make one of the conflicting parties change sides. You are faced with conditions that can be divided into rival groups, like two armies facing each other, and you must stop the war by finding enough deserters who will cross over.

You will remember that we used earlier the illustration of the evolving complexity and efficiency of cylinder head design. One of the most famous designs of the post-war years was that of the 4.2-litre Jaguar engine, shown in Fig. 20. Its efficiency sprang from the use of a spherical cylinder head, with inclined valves and two overhead camshafts.

After many years of successful production Jaguars

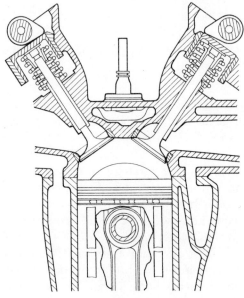

Fig. 20

decided to replace it with a multi-cylinder design where the costs for the individual cylinders had to be drastically reduced. In particular, they couldn't afford domed cylinder heads, inclined valves and two camshafts. In short, the features that made the first engine so efficient were precisely those that had to be discarded in the second, preferably without appreciable loss of efficiency. Three pairs of contradictions faced each other.

The solution to this conflict of interests begins with standing on one set of conditions and glaring critically at the opposing ones, then changing sides and looking back at the first list, always searching for some condition that you can persuade to come over to the enemy.

66

First let's start with the new conditions and look at the old engine. The double camshaft is almost essential for inclined valves. And inclined valves for domed cylinder heads, and domed combustion space, are essential for efficiency.

It is the domed shape which is the cause of most of the trouble; can we alter it to a flat one and so line it up on the economy side? It might bring over some of its companions with it. Go on looking at the confrontation and soon it will occur to you that the gases in the cylinder head don't really know or care whether the domed bit is on the top or the bottom, once the valves are shut they're too busy running round in circles. So why not put the dome in the top of the piston, since the gases won't notice the difference if no one tells them. Now the cylinder head can change sides. It is a flat face easily machineable, and over come the valves too, they can now be parallel and vertical, and so the double camshaft changes into one. And that is what Jaguars did – have a look at Fig. 21.

In this illustration the domed area quite literally changed sides, changed from the head to the piston, but this literal meaning is not normally essential or very likely. It is more an attitude of mind, open to all suggestions, even the apparently absurd, constantly probing each condition in turn seeking for the one clue that will resolve the conflict.

And this is not the only illustration of this that we can find in automobile engineering; the evolution of braking systems provides another. As the original drum brakes became increasingly effective they became increasingly hotter. As a result the drum expanded and distorted and edged away from the brake shoes. The more efficient it

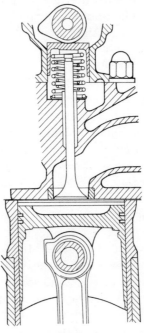

Fig. 21

became the less efficient it tried to be. And so the disc brake arrived and revolutionised the scene, and did it by making the expansion change sides. The hotter the disc became the more it expanded back at the brake pads. The civil war was over.

And looking to the future rather than the past we might be able to forecast how reinforced concrete columns may develop. In the orthodox design the steel is embedded in the concrete and when the load is applied they quarrel with each other about who is going to carry it. The concrete tends to creep and shorten itself so that

68

it can transfer more of its load to the steel, which com-
presses and hands some back. They are always trying to
pass the buck. But if, as has already been done on
occasions, you put the concrete inside the steel instead
of vice versa, the position becomes more amicable.
Sheathed in a steel tube the concrete cannot creep; there
is nowhere for it to creep off to. All it can do is to act
hydrostatically and thus re-inforce the tube against
buckling inwards. The creep has changed sides.

The third way to circumvent the confrontation of
conflicting interests is to outflank them.

This definition, like all the others, sounds vague and
ambiguous, and this is, perhaps, what you might well
expect, for words are not the natural language of
engineers. Drawings are their prose, mathematics their
grammar, and differential equations their poetry. Mere
words are inarticulate by comparison. And so we will
turn from the general to the particular – to a particular
design illustration. A large and long machine has been
devised for washing and drying materials, by squirting
them with liquids, squeezing and heating them. The
materials are fed through the machine sandwiched
between two conveyors, which must hold them flat but
impede the cleaning process as little as possible.

Experiments show that fine nylon thread is probably
the best material for constructing these conveyor bands
but, even so, they wear quite rapidly and so must be
easily replaceable.

And here it is that problems come piling on top of
each other. The conveyor bands must be renewable
without taking the machine to bits, be easily tensioned
when running, and be laterally stable.

To cut a long story short you end up with two possible,

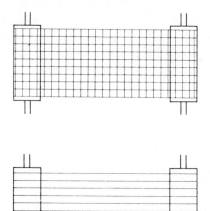

Fig. 22

or, more accurately, impossible designs, shown in Fig. 22.

The first is a nylon net. To put it on you have to feed it through the machine and then join all the ends up, and there are hundreds of these. Even worse, when you apply tension it tries to form a 'waist' in the middle and so you must fit countless little wheels and runners at the sides to keep it laterally stable. It would drive a maintenance engineer mad.

The second design uses each nylon thread as an individual conveyor belt, mutually spaced by combs (not shown).

This is laterally stable, but each little nylon thread tends to run at a different tension and all must still be joined up individually. The whole future of the machine

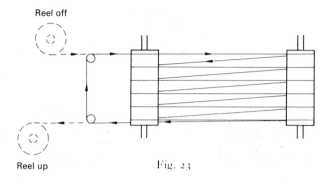

Reel off

Reel up Fig. 23

may depend on your finding a suitable solution; certainly, these two are no good. Well, outflank them. Go round
the edge somehow. By-pass the impasse. And you can
do it quite easily, as in Fig. 23. Here a single continuous
thread is used by making it angle slightly on the non-
conveying side. The tension stabilises itself throughout
and can be easily controlled. To replace it all, stop the
machine for a moment, break the thread at one end and
knot it to a fresh reel of nylon, and be ready to wind off
the old thread on to another reel (both shown dashed).
Start the machine up and the thread will progressively
be replaced as the machine runs. When all the worn
nylon is safely wound back on to one reel and the new
thread is in position, stop the machine momentarily.
Break the two threads near their reels, knot them
together, and start up again. The down time is reduced
from many hours to a few minutes.

Of course this outflanking solution should never have
been necessary, the design should have been in the
original list of possibilities. And 'Is there any way of
being certain that you *have* included all the possibilities?'

71

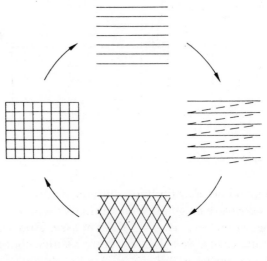

Fig. 24. A closed loop of conveyor design

is a question that obviously now arises. There is one, I think, and that is when you can take your list and logically join its two ends up. If you are inside a circle you cannot outflank it, there is no corner to edge past.

In the case of these nylon conveyors we could have completed such a circle of possibilities, as shown in Fig. 24.

The final two methods of tackling these situations of deadlock are even more difficult to define as they are more attitudes of mind than design techniques. The first comes into play when the apparent impossibility turns out to be a real one. And then you must accept the impossibility and not try to fight it. If a thing cannot be done, well, it cannot, and you had better face up to it.

You may remember that we mentioned more than

once the part sub-conscious thinking plays in creative design. And to exploit its potential you must first concentrate your conscious mind on the problem to be solved. Really concentrate; and you will never do this if half your mind is still quarrelling with the facts. You must put your whole mind to it. Now there are two types of impossibility which are always bona fide ones. And it is no good fighting them; they will win anyway. The first is a 'physical' impossibility.

Some years ago a plastic material was being made on a machine where the critical point of the process lay in the final squeezing between two rolls. The designed effect could only be obtained by maintaining certain critical conditions in the nip of the rolls. One of the most critical was that of temperature difference. The bottom roll was kept at a maximum of 80°C, and the top roll had to have a surface temperature of 2°C. The main difficulty was that any water vapour around immediately condensed on the surface of the cold roll, formed water, and ruined the material.

The atmosphere in the factory was far from ideal in more ways than one. There appeared to be a running fight between the various operators over how to find compromise temperatures; everyone was intent on 'proving' their ideas to be right and the other people's wrong – and sometimes leaving the door open to the British weather 'by mistake' just to help. Any views from an outsider risked his head being bitten off. This was discouraging as I was the outsider imported with the urgent instruction to 'stop the water condensing on the cold roll'. And naturally, this statement was, quite literally, a physical impossibility: water vapour condenses on a cold surface when it wants to and you cannot

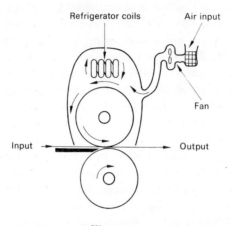

Fig. 25

negotiate a special agreement with the physical universe to make your factory an exception.

So I said to myself: 'Well, if it wants to condense I must accept the fact and let it get on with it. Not only that, but I will encourage it to do it.' So I covered the cold roll with a hood, as in Fig. 25, and circulated the air within it past refrigerated cooling pipes at − 10 °C. Any moisture instantly condensed on these and lost all interest in the cold roll. A small fan drew in sufficient new air, through a water-removing chemical, to keep a slight dry draught whistling round the gap where the material was fed in. Peace was declared all round.

The second type of impossibility that should demand our respect is a process one. For instance, you can't make omelettes without breaking eggs.

I well remember a particular impossibility of this type that I fruitlessly battled with for some time. I was responsible for the design and production efficiency of a

74

very large hot-air heating system for drying vast quanti-
ties of wood chips. Their water content as they arrived
at the machine varied from 12 per cent to 45 per cent by
weight.

The chief problem was that of continuously adjusting
the rate of drying so that all the chips came out at the
other end with their water content reduced to 10 per
cent with a limit of $\pm \frac{1}{2}$ per cent. I had a design for doing
the necessary adjusting of the heating air temperature
automatically, but I was not sure if it was going to work
properly. It would be wise to have something up my
sleeve in case it did not.

There seemed to be only one practicable scheme for
this and that was to plough off continuously a sample
stream of chips from the main input conveyor, measure
their percentage water content, assume they were
typical of the main stream, and adjust the heating to
suit. Now the sample stream of chips, arriving on a small
conveyor, presented a highly uneven appearance: a
miniature mountain range of valleys and peaks, the
volume per unit length never being the same even for a
few millimetres.

This volume variation was complicated by the fact
that the entrained air amongst the chips was varying
continuously as well and produced a three-to-one range
in bulk density.

Nor was this the end of my troubles, because there
was a 50 per cent difference in the dry density of
individual chips according to the type of wood they
came from. And then, of course, there was the four-to-
one variation of water content within a chip itself.

I was, therefore, faced with an indeterminate flow of
material constantly at the mercy of four widely and

THE SELECTION OF DESIGN

wildly fluctuating variables and I had to find the ratio
between two of them without the others muddling
everything up; and do it continuously; and do it
accurately; and find a mean value over short periods to
which the calibration of the dryer could be linked. Many
an hour I spent fighting those four variables. No sooner
than I tied up one, others would escape.

If I rolled out the bulk density, I was no nearer
capturing the others.

A complicated device producing a fixed rate of mass
flow still left me unable to distinguish between a light
wood with much water in it and a heavier wood with
less, and so on.

I only won the battle by giving up fighting it.

If these four variables wanted to come charging down
the conveyor, well let them. Don't interfere. Don't try
to roll them flat or anything else, but as they go by keep
a continuous record of the weight they exert on the
conveyor over a short length, by putting a weighing pan
underneath. They won't know it is happening. Now let
them pass under an intense induction-type heater which
will drive off the moisture in seconds. Next let them ride
over another weighing pan, beneath the conveyor,
which also takes continuous readings. Fig. 26 shows the
general layout.

You can easily arrange for the information from the
first weighing pan to be electrically delayed for exactly
the same period as it takes for a chip on the conveyor
to be carried from the first weighing pan to the second.

Then compare continuously the weight on the second
one (the dry weight) with the delayed reading on the
first which will give you the mass of the same set of
chips when they were wet. The instantaneous and

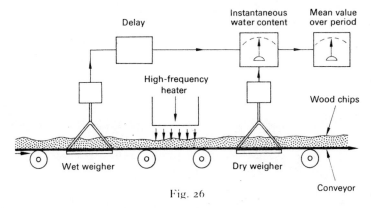

Fig. 26

running mean moisture contents are now yours for the asking. Don't fight the impossible but juggle with the possible.

Lastly there is a technique that can only, I think, be understood by experience. To call it an 'extra dimension of thinking', or possibly 'the exploitation of the unexpected' is not particularly helpful, but having done it you can look back on these definitions and see a certain element of truth in them.

77

7

A DETECTIVE STORY

How are you getting along with your substitute for wood?

'I thought you had forgotten all about that.'

I haven't and I hope you haven't either.

'Actually, I have given it some more thought while you have been going on and on but I still don't see how I can consolidate all those different bits together when they are all wet and sticky. You might be able to do it by hand; by a machine, never.'

You are up against a process impossibility, in fact?

'Yes, I suppose so.'

Well, if you can't beat them join them. If the only possible way of putting it all together is by leaving the water out, accept the fact and see where it leads you.

'If I had a box made of flat steel plates, whose inside was the exact shape of a wooden beam or slab, I could then take one edge off and drop the materials in their relative positions from hoppers, as in Fig. 27. The air passages would be formed by tubes with their lower ends fixed to the bottom edge of the box. I would probably have to use vibration to help the materials in and even then I might have air voids.'

How could you get rid of those?

'I suppose I would have to squeeze it together; I'm not sure really.'

Squeezing it would be frightfully complicated, can't you make it all expand and so squeeze itself?

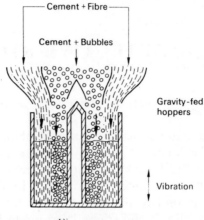

Fig. 27

'I see what you're getting at; heat it all up in an oven and the polystyrene air bubbles will expand and make a pressure.'

Yes, but I wouldn't necessarily use an oven; there are other ways of heating things.

'What do I try next?'

Well, you have achieved your first step; you have, in a fairly amicable manner, persuaded your dry materials to settle into their appropriate positions. Now try to get the water injected somehow.

So far you have been borrowing orthodox concrete consolidating techniques and exploiting them to a particular end, and I think that this is as far as you can go in this way. Now you must try and widen your horizon, forget the orthodoxy, collect clues and see if they can suggest some new medium to exploit.

'Clues make it sound like a detective story.'

In some ways an engineering design *is* like a detective

story, especially when they are both good ones. There
are usually a number of clues scattered through the
murder mystery and they all point to the murderer, but
you only learn this in the final chapter. Then it all seems
so obvious that you kick yourself for not seeing it all
sooner.

And the climax of an engineering design often lies in
the unexpected revelation that co-ordinates the clues
and solves the problem, and you feel you should have
spotted it all earlier.

'Are you going to tell me how to conjure water into an
enclosed box and force it through a mass of bits and
grit?'

No, you're going to tell me. Take your blinkers off,
concentrate on your clues and you should be able to see
the plot.

'Can't you tell me what the clues are?'

Haven't you spotted them? I've been pelting you
with them. 'Heat', 'pressure', 'force', 'water' and
'inject' are some of them.

'I still can't imagine what you are driving at.'

Go away and make yourself a cup of tea; relax and let
your subconscious mind go to work . . .

'. . . I'm back, but it wasn't any good; I still can't see
a solution. I prefer coffee to tea anyway.'

Perhaps making tea would have helped. If you gaze
at a kettle and mutter 'heat', 'pressure', 'force' and
'water' to yourself, mightn't something occur to you?

'Yes, of course, sorry I've been so dense. Put those
four words together and they stand for steam. But how
do I get the steam in?'

Make those pipes porous, and feed steam in at the
bottom of them. It will force itself out through the

porous sides of the pipes uniformly, meet the cool cement and condense into boiling water, which will spread out through the cement, heating up the bubbles and expanding them as it passes.

'Yes, I can see it all in principle, but there are a number of details that need working out.'

One of my jobs at the moment is to do just that and I'm not going to give you any more hints, you might beat me to it. If you happen to try I must warn you that there are lots of patents about that will make it awfully awkward for you.

But there are plenty of other things to invent; go off and see what you can do; few other professions can equal the adventure of inventiveness, especially if you can pioneer something that is an asset to the community, for you can enjoy watching other people enjoying what you have given them.

INDEX